Kaladhar DSVGK

Studies on antimicrobial, biochemical and image analysis in Mirabilis jalapa

Anchor Compact

DSVGK, Kaladhar: Studies on antimicrobial, biochemical and image analysis in Mirabilis jalapa, Hamburg, Anchor Academic Publishing 2014

Buch-ISBN: 978-3-95489-301-0
PDF-eBook-ISBN: 978-3-95489-801-5
Druck/Herstellung: Anchor Academic Publishing, Hamburg, 2014

Bibliografische Information der Deutschen Nationalbibliothek:
Die Deutsche Nationalbibliothek verzeichnet diese Publikation in der Deutschen
Nationalbibliografie; detaillierte bibliografische Daten sind im Internet über
http://dnb.d-nb.de abrufbar

Bibliographical Information of the German National Library:
The German National Library lists this publication in the German National Bibliography.
Detailed bibliographic data can be found at: http://dnb.d-nb.de

© Anchor Academic Publishing, ein Imprint der Diplomica® Verlag GmbH
http://www.diplom.de, Hamburg 2014
Printed in Germany

INDEX

INTRODUCTION

Mirabilis jalapa Linn. (Family: Nyctaginaceae) is a widely used in conventional medicine in many parts of the world for the treatment of various diseases viz. virus inhibitory activity, anti-tumor activity, etc. Very few reports are available on Architecture of pollen grains, image analysis, Antimicrobial activity, pharmacognostic and phytochemical nature of *Mirabilis jalapa* Linn. The genus Mirabilis contains 350 species in 34 classifications. The common garden-variety four-o'clock (*Mirabilis jalapa*) is also known as Marvels of Peru. Four o'clock received its name because of its habit of opening in the late afternoon. It is not actually the time of day that causes the flowers to open, but the drop in temperature. The flowers close the next morning, except on dull, cloudy days.

An interested aspect of this plant is that flowers of different colors can be found simultaneously on the same plant. Additionally, an individual flower can be splashed with different colors. Another interesting point is a color-changing phenomenon. For example, in the yellow variety, as the plant matures, it can display flowers that gradually change to a dark pink color. Similarly white flowers can change to light violet.

Mirabilis jalapa (**Plate 1**) is said to have been exported from the Peruvian Andes in 1540. Around 1900, Carl Correns used the four o'clock as a model organism for his studies on cytoplasmic inheritance. The flowers are used in food colouring. An edible crimson dye is obtained from the flowers to colour cakes and jellies. In herbal medicine, parts of the plant may be used as a diuretic, purgative, and for vulnerary (wound healing) purposes. The root is believed an aphrodisiac as well as diuretic and purgative. It is used in the treatment of dropsy. Powdered, the seed of some varieties is used as a cosmetic and a dye. The seeds are considered as poisonous can cause vomiting and diarrhea and, in large quantities, death.

Each plant may have hundreds of two-inch, trumpet-shaped blossoms that will last throughout the summer until the first frost. Even on a single plant, these blossoms may be pink, yellow, white, salmon, red, or striped or blotted with any of these colors. Because of their variety of shades, though, they are hard to co-ordinate with other garden flowers and are best used as bedding, border, background, or in containers.

In mining, geology, and quarry production, it is well known that the properties of aggregates, such as size and shape, are very important information for particle characterization and the optimization of production. For the last fifteen years, image analysis techniques have been used for aggregate particle measurement, which increases speed and accuracy of analysis. There are a number of methods for measuring aggregate particle size and shape in image analysis, the stability of measurement methods is very important.

Palynology is the study of plant pollen, spores and certain microscopic plankton organisms (collectively termed palynomorphs) in both living and fossil form. Botanists use living pollen and spores (actuopalynology) in the study of plant relationships and evolution, while geologists (palynologists) may use fossil pollen and spores (paleopalynology) to study past environments, stratigraphy (the analysis of strata or layered rock) historical geology and paleontology.

Melissopalynology is the study of pollen in honey, with the purpose of identifying the source plants used by bees in the production of honey. This is important to honey producers because honey produced by pollen and nectar from certain plants as mesquite, buckwheat, tupelo or citrus trees demand a higher price on the market than that produced by other plant sources. Some plants may produce nectar and pollen that is harmful to human health. A careful monitoring of the pollen types found in honey may identify these toxic sources and the honey produced may be kept out of the commercial market.

Pollen grains are showing two nuclei. The smaller (generative) nucleus forms two sperm cells. Endoplasmic reticulum threads present throughout the pollen grain and several mitochondria and plastids containing starch.

PLATE 1

Fig: 1. <u>Mother Plants</u>

A. Yellow variety   B. Pink variety   C. Orange variety

Fig: 2. <u>Mother Plants with Scale Reading</u>

3

OBJECTIVES

Mirabilis jalapa Linn. belongs to the family Nyctaginaceae and is a large herbaceous plant grown in gardens throughout India. The present project work entitled **"Studies on antimicrobial, biochemical and image analysis in *Mirabilis jalapa* varities"** contains following objectives:

1. Staining of pollen grain and observation of meiotic stages.
2. Phytochemical studies of 3 different colored plant varieties of *M.jalapa* by TLC
3. Comparative Evaluation of Antimicrobial Activities of plant leaf Extract (3 different colored varieties) of *Mirabilis jalapa*.
4. Image analysis of data got by experimentation using software.

REVIEW OF LITERATURE

Nature has been a source of medicinal agents for thousands of years and an impressive number of modern drugs have been isolated from natural sources, many based on their use in traditional medicine. Various medicinal plants have been used for years in daily life to treat disease all over the world. They have been used as a source of medicine. The widespread use of herbal remedies and healthcare preparations, such as those described in ancient texts like the Vedas and the Bible, has been traced to the occurrence of natural products with medicinal properties. In fact, plants produce a diverse range of bioactive molecules, making them a rich source of different types of medicines. Higher plants, as sources of medicinal compounds, have continued to play a dominant role in the maintenance of human health since ancient times [1]. Over 50% of all modern clinical drugs are of natural product origin [2] and natural products play an important role in drug development programs in the pharmaceutical industry [3].

There has been a revival of interest in herbal medicines. This is due to increased awareness of the limited ability of synthetic pharmaceutical products to control major diseases and the need to discover new molecular structures as lead compounds from the plant kingdom. Plants are the basic source of knowledge of modern medicine. The basic molecular and active structures for synthetic fields are provided by rich natural sources. This burgeoning worldwide interest in medicinal plants reflects recognition of the validity of many traditional claims regarding the value of natural products in health care.

The relatively lower incidence of adverse reactions to plant preparations compared to modern conventional pharmaceuticals, coupled with their reduced cost, is encouraging both the consuming public and national health care institutions to consider plant medicines as alternatives to synthetic drugs. Plants with possible antimicrobial activity should be tested against an appropriate microbial model to confirm the activity and to ascertain the parameters associated with it. The effects of plant extracts on bacteria have been studied by a very large number of researchers in different parts of the world [4-6]. Much work has been done on ethnomedicinal plants in India [7-9]. Interest in a large number of traditional natural products has increased [10]. It has been suggested that aqueous and ethanolic extracts from plants used in allopathic medicine are potential sources of antiviral, antitumoral and antimicrobial agents [11, 12]. The selection of crude plant extracts for screening programs has the potential of being more successful in initial steps than the screening of pure compounds isolated from natural products [13].

This plant is 50-100 cm high. It has antifungal, antimicrobial, antiviral, antispasmodic, antibacterial, diuretic, carminative, cathartic, hydragogues, purgative, stomachic, tonic and vermifuge properties. [14] This plant contains alanine, alphaamyrins, arabinose, beta-amyrins, campesterol, daucosterol and dopamine [15], and is used to treat conjunctivitis, edema, fungal infections, inflammation, pains and swellings.

Mirabilis jalapa L. (Nyctaginaceae) is a tropical American herb that is commonly cultivated in North America where it is perennial in the south and warm west and annual in the north. In *Mirabilis jalapa* pollen performance was influenced by the number of competing pollen grains or pollen tubes, but was not influenced by potential genetic differences with load diversity [16].

B P Cammue *et al.,* has isolated from seeds of *Mirabilis jalapa* L. two antimicrobial peptides, designated Mj-AMP1 and Mj-AMP2, respectively. These peptides were also active on two tested Gram-positive bacteria but were apparently nontoxic for Gram-negative bacteria and cultured human cells [17].

J Kataoka *et al.*, 1991, cloned a cDNA for Mirabilis antiviral protein (MAP), a ribosome-inactivating protein (RIP), which inhibits the mechanical transmission of plant virus and the in vitro protein synthesis of both prokaryotes and eukaryotes [18]. A potent antiviral activity was found in extracts from a yellow flower cultivar of *Mirabilis jalapa* L. (Nyctaginaceae) in root, leaf, and stem tissues and in *in vitro* cultured cells [19,20].

Extracts of *Mirabilis jalapa* (Nyctaginaceae), containing a ribosome inactivating protein (RIP) called Mirabilis antiviral protein (MAP), were tested against infection by potato virus X, potato virus Y, potato leaf roll virus, and potato spindle tuber viroid. Several plants, such as *Pelargonium hortorum, Chenopodium album, C. amaranticolor, Capsicum frutescens, Azadirachta indica, Vitis vinifera,* and *Rosa banktia*, possess antiviral factors. Plant-derived antiviral compounds are active against plant, animal, and human viruses. Plant antiviral compounds are grouped as furocoumarins, alkaloids, terpenoids, lignins, and specific proteins. Among plant-derived antiviral proteins, a group called ribosome-inactivating proteins (RIPs), which are widely distributed in higher plants, hold promise for agricultural and pharmaceutical applications [21].

MATERIALS AND METHODS

Study species

Mirabilis jalapa has tubular flowers are fragrant and vary in color among plants. The self-compatible, perfect flowers each have 5–6 stamens and a single-ovulate ovary. An individual flower opens for one night in the early evening, the exact time depending on temperature and relative humidity, and closes early the next morning. An individual plant produces between 25 and 75 flowers in one flowering season.

Growth form: Annual or perennial herb.

Size: 0.5 - 2 m tall.

Root : a swollen and somewhat tuberous taproot.

Stems: usually several, erect to slightly decumbent, branching, light or bright green, sometimes with a yellow or pink hue, mostly hairless, but sometimes hairy near the base or even glandular-hairy further up; if hairy, often in two lines.

Leaves: about midstem and above, opposite, on 1 - 7 cm long stalks, somewhat elongate triangular to egg-shaped or lance-shaped, 4 - 14 cm long, 2 - 9 cm wide, with blunt or indented bases, and non-toothed edges.

Inflorescence: of several, terminal or axillary, compact clusters with one to fifteen flowers on short, 0.5 - 5 mm long stalks, and each cluster subtended by a pair of 2 - 17 mm long leaf-like bracts. Each flower sits atop a green, 0.5 - 1.5 cm tall, bell-shaped cup (involucre) formed by five fused bracts with triangular tips.

Flowers: pink or yellow or orang, sometimes white or striped, usually hairless, 3 - 5 cm long, radially symmetric, funnel-shaped with a long, narrow tube, and five, abruptly flared lobes.

Sepals: showy, brightly colored, not green, and mimicking petals. The five sepals are fused for most of their length, constricted above the ovary into a long, narrow tube, then separated into five, abruptly flared lobes.

Petals: none.

Stamens: five, long, and extending beyond the sepal tube.

Pistil: with one, single-chambered, superior, somewhat globular ovary; one, long, threadlike style, which extends beyond the stamens; and a rounded, head-like stigma.

Fruit: a hard, dark brown or nearly black, 0.7 - 1.1 cm long, broadly ellipsoid to inversely egg-shaped, one-seeded, nut-like achene, which is tightly enclosed by the remnant bract cup (involucre). The achene is round or slightly five-angled in cross-section, smooth or inconspicuously bumpy or warty, and either hairless or hairy.

Plant Materials

M. jalapa (yellow, orange and pink flower cultivars) were collected from plants grown in the garden of the GITAM Institute of sciences, GITAM University, Visakhapatnam during winter season.

Meiosis in flower buds of *Mirabilis jalapa*

Materials: *M.jalapa* flower bud, Tween 80, acetocarmine stain, glass slides, cover slips, blotting paper.

Procedure:

1. Select appropriate flower buds of different size from the inflorescence.

2. Fix them in Tween 80 fluid, which is used as fixative.

3. Take a preserved flower bud and place it on a glass slide.

4. Separate the anthers and discard the other parts of the bud.

5. Put one or two drops of acetocarmine and squash the anthers.

6. Leave the material in the stain for five minutes.

7. Place a cover slip over them and tap it gently with a needle or pencil.

8. Warm it slightly over the flame of a spirit lamp.

9. Put a piece of blotting paper on the cover slip and apply uniform pressure with the thumb.

10. Observe the slide under the light microscope for different pollen grain meiotic stages at 10X, 45X and 100X.

Image Analyser v 1.31

The GSA Image Analyser v 1.21 (2009) is a program for the scientific evaluation of 2D images (image analysis). The possibilities of the program are varied and can be divided in three main groups:

- Object surface calculation
- Object length calculation
- Object counting

Surface calculation

The surfaces of all recognized objects are automatically calculated. In addition, the surface calculation of single objects is possible. The program offers the possibility to calculate the relations of the ascertained surfaces to each other and to the whole image or background.

Object length calculation

The GSA Image Analyser has a set of functions which have been developed especially for the direct length calculation of very complex and much ramified objects. In addition to the direct length calculation, an indirect calculation with a grid intersection (developed by Tennant) is possible. For this function, the grid size can be freely defined.

Object counting

The automatic count-function determines the number of all recognised objects and, in addition, offers the possibility to count overlapping objects apart. This separation can make over the object form, the object surface and over the object intensity.

The grid-intersection-count-function serves for the indirect object number regulation. Such procedures are described, e.g., by Buerker, Fuchs-Rosenthal, Thoma, shilling, Tuerk.

Beside the automatic count-functions, the program has two manual count-procedures included. One method marks the objects inside of the image and the other cuts the image into several smaller parts to make it easier for human eyes to follow many small objects.

Picture import

The GSA Image Analyser can read almost all known picture formats (JEPEG, GIF, TIFF, BMP, PNG). In addition, the program provides interfaces to scanners, microscopes and digital cameras as well as to video cameras (Video grabber, TV cams).

Picture creation and manipulation

A picture production by means of input device (graphics tray, mouse) is integrated in the program and allows a cloning of objects. In addition, there exist picture manipulation tools and different filter functions.

Calibration

This function allows calibrating the input tools as well as determining the picture resolution (DPI) with picture formats who do not contain this information.

This program is designed to solve many problems. It has been used for counting cells and calculating the size of plant.

Image Analyzer is a robust, proprietary, real time image analysis system. The engine is supplied with an API, is external dependency independent and supports multiple platforms (Windows/LINUX/Unix etc). It is small in footprint, fast and accurate when identifying pornographic or inappropriate image content within all popular digital data transmission protocols.

Anti microbial activity

Sample extraction

The determined Fresh plant leaves (200g) were ground, extracted with Diethyl ether, ethyl acetate and methanol separately and filtered. The plant residue was re-extracted by adding above solvents and filtered again after 48hs. Such procedure was repeated every 72hs, completing three filtration processes. The filtrate was concentrated on a rotary evaporator at 45°C for solvent elimination, and the extracts were kept in sterile bottles under refrigerated conditions until use. The final volume is adjusted to a concentration of 1 mg/ml with the above solvents separately.

Test microorganisms

The microbial strains are identified strains and were obtained from the **MTCC**, IMTECH, Chandigarh, India. The bacterial strains studied are *Eschericia coli* **MTCC** 739, *Staphylococcus aureus* **MTCC** 737, *Klebsiella pneumoniae* **MTCC** 109, *Bacillus subtilis* **MTCC** 441 and *Aspergillus niger* **MTCC** 282.

Antimicrobial assay

The antibacterial assays were performed by the agar well diffusion method. Petri dishes (200 mm) were poured with nutrient agar (HI-Media) and allowed to solidify to make base layers. The seed layers were prepared by inoculating 10mL of test organism suspension in 100 mL Mueller-Hinton agar(for bacteria) and Sabouraud Dextrose agar (for fungi) and wells, 6 mm in diameter, were made in the agar medium with the help of a sterile steel borer. About 100 µL of each extract was added aseptically in wells. All the plates were incubated at $37 \pm 1°C$ for 24 hours in the upright position. At the end of the incubation times, the diameters of the inhibition zones were measured in millimeters. Ampicillin (20 µg/mL) and solvent (diethyl ether, ethyl acetate, and methanol 20 µg/mL) were used as antimicrobial compounds against text microorganisms. The tests were conducted in triplicate.

Thin layer chromatography (TLC)

Thin layer chromatography (TLC) is a chromatography technique used to separate mixtures. The present study is performed on a sheet of glass, which is coated with a thin layer of adsorbent material, usually silica gel. This layer of adsorbent is known as the stationary phase.

After the sample has been applied on the plate, a solvent or solvent mixture (known as the mobile phase) is drawn up the plate via capillary action. Because different analytes ascend the TLC plate at different rates, separation is achieved.

The TLC analysis was performed on glass slides pre-coated with silica gel G/GF (E-Merck grade). Before use, glass slides were pre-washed with methanol, and dried in an oven at 105°C for 1 hour.

Plates are prepared with 5g Silica Gel G and GF in a ratio of 8:2 by using 5ml ethyl acetate as a solvent.

A small spot of solution containing the sample is applied to a plate, about one centimeter from the base. The plate is then dipped in to a suitable solvent, such as hexane or ethyl acetate, and placed in a sealed container. The extracts (10 µL) were applied on the plates as bands of 7-mm width with the help of a linomat-5 sample applicator. The solvent moves up the plate by capillary action and meets the sample mixture, which is dissolved and is carried up the plate by the solvent. Different compounds in the sample mixture travel at different rates due to the differences in their attraction to the stationary phase, and because of differences in solubility in the solvent. By changing the solvent, or perhaps using a mixture, the separation of components (measured by the R_f value) can be adjusted. TLC plate is visualized under UV radiation.

TLC runned in ethylacetate and hexane in the ratio of 1:9, 3:7 and 5:5 for 10%,30% and 50% mobile phase.

Image analysis

Image analysis is done on Meiotic stages using **Pixcavator IA - Image Analysis 2.4.1.**

Pixcavator's operation is dead simple. Push "Run" and in a few seconds the software displays a list of all objects in the image. The list gives the size, location, and other characteristics of each object.

The areas of applications of Pixcavator are unlimited: medical and ecological, microscopy and satellite.

Pixcavator IA - Image Analysis 2.4.1 image analysis methods are based on a rock solid mathematics. This mathematics has been developed in the last 100 years but only recently its methods became useful in practical applications.

The underlying idea is very simple. The "objects" captured by the software are connected clusters of darker pixels surrounded by lighter background, or vice versa. For example, in the image below, the gray cells are captured as "objects" because they are surrounded by black pixels.

Counting:

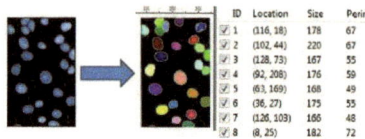

Measuring:

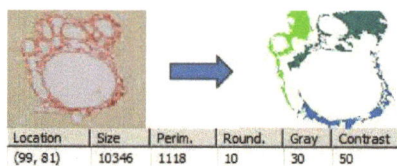

Once these objects are captured, their sizes are measured, their locations are found, and their shapes are evaluated. The results are presented to the user.

Major Features:

1. Digital image processing, analysis, and manipulation; elementary computer vision.
2. Automatic, semi-automatic, and manual analysis.
3. Pixcavator provides new image analysis capabilities to scientists and researchers.
4. With a single click you can capture the contours of all objects in the image and produce a spreadsheet with each objects locations and measurements.
5. You can mark objects in the spreadsheet to highlight objects in the image, or vice versa.
6. Extract or remove objects as desired.
7. The software also includes all standard image processing tools.

Camera used: Olympus MODEL- FE-115, 5 megepixel camara (2048X1536).

RESULTS

Mirabilis jalapa is a vulnerable medicinal plant, having wide importance based on our research. The plant is widely available in and around the regions of Visakhapatnam district. There is a huge varieties and characteristic features in this plant and can be mutated easily due to the structural arrangement of pollen grain.

Staining of pollen grain

Plate 2 provides the pollen grain is the male gametophyte generation of *M.jalapa*. Mass of microscopic spores in a seed plant that appears usually as a fine dust is a pollen. Each pollen grain is tiny, varies in shape and structure. The anther is a 2 lobed structure attached to the stamen and is further subdividing into 7 sub lobes. The outer layer of the anther is thick, middle cellular and inner slimmer layers. The middle layer which is cellular may be represented as anther and the number is many (Fig 1). Dr.Robert marcus, from Biological Research Center of the Hungary has already been researched on pollen grains of *M. jalapa* and is image is shown in fig:2

Fig:3. Has provide the architecture of pollen grain visualized at 100x microscopy. The architecture is some what complex and has been adopted based on climatic factors. The surface of mature pollen grain is having pores which are arranged in serial fashion at half posterior and axial another anterior. The posterior half has pores arranging from 1 to 5. The anterior portion ranges 1, 3, 5 on first half and another half. (Fig.6). The anterior side is based on the radial arrangement may contain nucleolus and DNA due to its adaptation. The large number of pores on the grain may be adopted for pollination where the sticky spores attach to the insects and will move from anther to stigma of the flower or neighboring flower, where fertilization takes place.

Fig: 4 have provided the partial staining of pollen grain nucleous and nucleolus of the pollen grain and is clearly visualized in this fig. the DNA is condensed in the nucleolus.

Fig: 5 is well strained and various organelles of pollen grain is visualized. The two outer layers of the pollen grains, exine and intine are clearly visible. The internal organelles such as endoplasmic reticulum, chromoplasts, mitochondria and DNA is visualized in this figure.

Plate 2

Fig: 1. Mirabilis jalapa stamen and anther

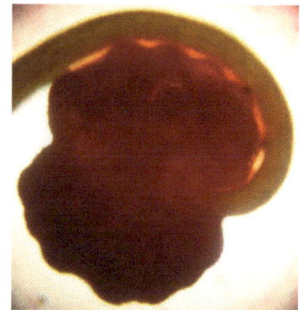

Fig: 2. Mirabilis jalapa stamen and pollen at 125x

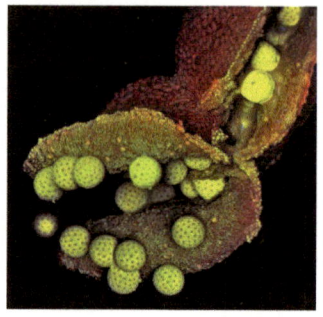

Fig:2 from Author: Dr. Robert Markus

Working place: Institute of Genetics

Biological Research Center of the Hungarian

Academy of Sciences

Szeged, Hungary

Fig:3. Pollen grain (without tween in Staining) at 100x

Fig:4. Pollen grain (with moderate staining) at 100x

Fig:5- Pollen grain using (stained) at 100x

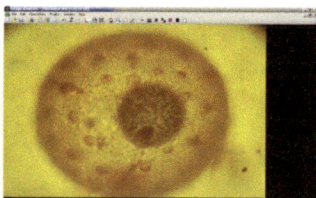
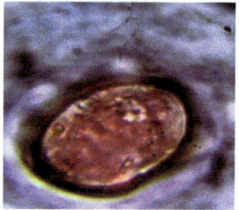

Fig.6 Architecture of pollen grain

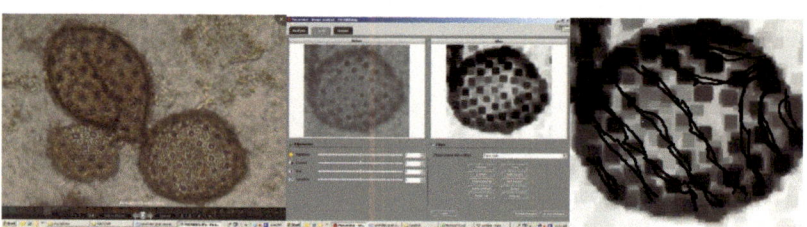

MEIOSIS

In biology, meiosis is a process of reductional division in which the number of chromosomes per cell is cut in half. Meiosis gives rise to spores. As with mitosis, before meiosis begins, the DNA in the original cell is replicated during S-phase of the cell cycle. Two cell divisions separate the replicated chromosomes into four haploid gametes or spores.

Meiosis is essential for sexual reproduction and therefore occurs in all eukaryotes (including single-celled organisms) that reproduce sexually. During meiosis, the genome of a diploid germ cell, which is composed of long segments of DNA packaged into chromosomes, undergoes DNA replication followed by two rounds of division, resulting in four haploid cells. Each of these cells contains one complete set of chromosomes, or half of the genetic content of the original cell. In all plants, and in many protists, meiosis results in the formation of haploid cells that can divide vegetatively without undergoing fertilization, referred to as spores. In these groups, gametes are produced by mitosis.

Plate 3 describes about meiosis I and then meiosis II and is imaged using image analyzer software.

Meiosis I consists of separating the pairs of homologous chromosome, each made up of two sister chromatids, into two cells. One entire haploid content of chromosomes is contained in each of the resulting daughter cells; the first meiotic division therefore reduces the ploidy of the original cell by a factor of 2.

Meiosis II consists of decoupling each chromosome's sister strands (chromatids), and segregating the individual chromatids into haploid daughter cells. The two cells resulting from meiosis I divide during meiosis II, creating 4 haploid daughter cells. Meiosis I and II are each divided into prophase, metaphase, anaphase, and telophase stages, similar in purpose to their analogous sub phases in the mitotic cell cycle. Therefore, meiosis includes the stages of meiosis I (prophase I, metaphase I, anaphase I, telophase I), and meiosis II (prophase II, metaphase II, anaphase II, telophase II).

Meiosis I

Meiosis I separates homologous chromosomes, producing two haploid cells, so meiosis I is referred to as a reductional division. This is because later, in Anaphase I, the sister chromatids will remain together as the spindle pulls the pair toward the pole of the new cell.

In meiosis II, an equational division similar to mitosis will occur whereby the sister chromatids are finally split, creating a total of 4 haploid cells per daughter cell from the first division.

Prophase I

During prophase I, DNA is exchanged between homologous chromosomes in a process called homologous recombination. This often results in chromosomal crossover. The new combinations of DNA created during crossover are a significant source of genetic variation, and may result in beneficial new combinations of alleles. The paired and replicated chromosomes are called bivalents or tetrads, which have two chromosomes and four chromatids, with one chromosome coming from each parent. At this stage, non-sister chromatids may cross-over at points called chiasmata (plural; singular chiasma).

Leptotene (Fig 2.1)

The first stage of prophase I is the leptotene stage, also known as leptonema, from Greek words meaning "thin threads". During this stage, individual chromosomes begin to condense into long strands within the nucleus. However the two sister chromatids are still so tightly bound that they are indistinguishable from one another.

Zygotene (Fig 2.2)

The zygotene stage, also known as zygonema, from Greek words meaning "paired threads", occurs as the chromosomes approximately line up with each other into homologous chromosomes. This is called the bouquet stage because of the way the telomeres cluster at one end of the nucleus. At this stage, the synapsis (pairing/coming together) of homologous chromosomes takes place.

Pachytene (Fig 2.3)

The pachytene stage, also known as pachynema, from Greek words meaning "thick threads" contains the following chromosomal crossover. Nonsister chromatids of homologous chromosomes randomly exchange segments of genetic information over regions of homology. Sex chromosomes, however, are not wholly identical, and only exchange information over a small region of homology. Exchange takes place at sites where recombination nodules (the chiasmata) have formed. The exchange of information between the non-sister chromatids results in a recombination of information; each chromosome has the complete set of information it had before, and there are no gaps formed as a result of the process. Because the chromosomes cannot be distinguished in the synaptonemal complex, the actual act of crossing over is not perceivable through the microscope.

Diplotene (Fig 2.4)

During the diplotene stage, also known as diplonema, from Greek words meaning "two threads", the synaptonemal complex degrades and homologous chromosomes separate from one another a little. The chromosomes themselves uncoil a bit, allowing some transcription of DNA. However, the homologous chromosomes of each bivalent remain tightly bound at chiasmata, the regions where crossing-over occurred. The chiasmata remain on the chromosomes until they are severed in Anaphase I.

Diakinesis (Fig 2.5)

Chromosomes condense further during the diakinesis stage, from Greek words meaning "moving through". This is the first point in meiosis where the four parts of the tetrads are actually visible. Sites of crossing over entangle together, effectively overlapping, making chiasmata clearly visible. Other than this observation, the rest of the stage closely resembles prometaphase of mitosis; the nucleoli disappear, the nuclear membrane disintegrates into vesicles, and the meiotic spindle begins to form.

Synchronous processes

During these stages, two centrosomes, containing a pair of centrioles in animal cells, migrate to the two poles of the cell. These centrosomes, which were duplicated during S-phase, function as microtubule organizing centers nucleating microtubules, which are essentially cellular ropes and poles. The microtubules invade the nuclear region after the nuclear envelope disintegrates, attaching to the chromosomes at the kinetochore. The kinetochore functions as a motor, pulling the chromosome along the attached microtubule toward the originating centriole, like a train on a track. There are four kinetochores on each tetrad, but the pair of kinetochores on each sister chromatid fuses and functions as a unit during meiosis I.

Microtubules that attach to the kinetochores are known as kinetochore microtubules. Other microtubules will interact with microtubules from the opposite centriole: these are called nonkinetochore microtubules or polar microtubules. A third type of microtubules, the aster microtubules, radiates from the centrosome into the cytoplasm or contacts components of the membrane skeleton.

Metaphase I (Fig 2.6)

Homologous pairs move together along the metaphase plate: As kinetochore microtubules from both centrioles attach to their respective kinetochores, the homologous chromosomes align along an equatorial plane that bisects the spindle, due to continuous counterbalancing forces exerted on the bivalents by the microtubules emanating from the two kinetochores of homologous chromosomes. The physical basis of the independent assortment of chromosomes is the random orientation of each bivalent along the metaphase plate, with respect to the orientation of the other bivalents along the same equatorial line.

Anaphase I (Fig 2.7)

Kinetochore microtubules shorten, severing the recombination nodules and pulling homologous chromosomes apart. Since each chromosome has only one functional unit of a pair of kinetochores, whole chromosomes are pulled toward opposing poles, forming two haploid sets. Each chromosome still contains a pair of sister chromatids. Nonkinetochore microtubules lengthen, pushing the centrioles farther apart. The cell elongates in preparation for division down the center.

Telophase I (Fig 2.8)

The last meiotic division effectively ends when the chromosomes arrive at the poles. Each daughter cell now has half the number of chromosomes but each chromosome consists of a pair of chromatids. The microtubules that make up the spindle network disappear, and a new nuclear membrane surrounds each haploid set. The chromosomes uncoil back into chromatin. Cytokinesis, the pinching of the cell membrane in animal cells or the formation of the cell wall in plant cells, occurs, completing the creation of two daughter cells. Sister chromatids remain attached during telophase I.

Cells may enter a period of rest known as interkinesis or interphase II. No DNA replication occurs during this stage.

Meiosis II

Meiosis II is the second part of the meiotic process. Much of the process is similar to mitosis. The four main steps of Meiosis II are: Prophase II, Metaphase II, Anaphase II, and Telophase II.

Fig 2.9 shows **Prophase II,** takes an inversely proportional time compared to prophase I. In this prophase we see the disappearance of the nucleoli and the nuclear envelope again as well as the shortening and thickening of the chromatids. Centrioles move to the polar regions and arrange spindle fibers for the second meiotic division.

Fig 2.10 metaphase II, the centromeres contain two kinetochores that attach to spindle fibers from the centrosomes (centrioles) at each pole. The new equatorial metaphase plate is rotated by 90 degrees when compared to meiosis I, perpendicular to the previous plate.

This is followed by anaphase II (Fig 2.11), where the centromeres are cleaved, allowing microtubules attached to the kinetochores to pull the sister chromatids apart. The sister chromatids by convention are now called sister chromosomes as they move toward opposing poles.

The process ends with telophase II (Fig 2.12), which is similar to telophase I, and is marked by uncoiling and lengthening of the chromosomes and the disappearance of the spindle. Nuclear envelopes reform and cleavage or cell wall formation eventually produces a total of four daughter cells, each with a haploid set of chromosomes. Meiosis is now complete and ends up with four new daughter cells.

PLATE 3

Fig: 1. Stained specimen under 100x objective

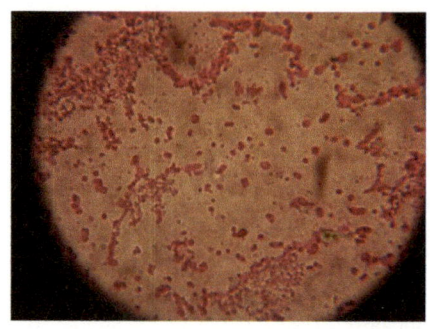

Fig: 2. Meiotic stages (Using Pixcavator 2.4.1)

Prophase I

2.1 Leptotene   2.2 Zygotene   2.3 Pachytene   2.4 Diplotene   2.5 Diakinesis

2.6 Metaphase I   2.7 Anaphase I   2.8 Telophase I

2.9 Prophase II   2.10 metaphase II   2.11 anaphase II   2.12 telophase II

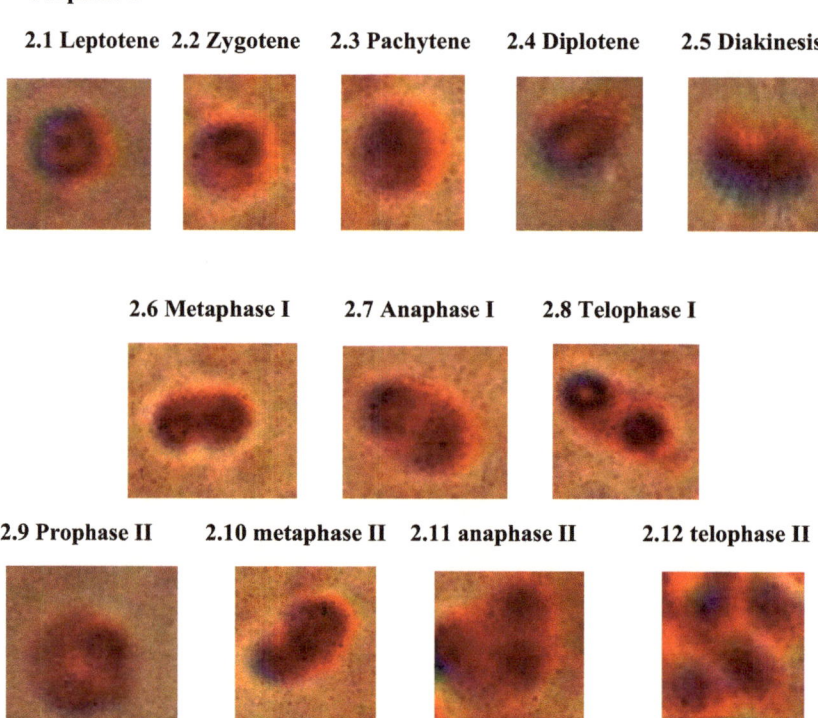

THIN LAYER CHROMATOGRAPHY (TLC)

TLC is an analytical method for isolation and characterization of various metabolites. Primary metabolites such as proteins, carbohydrates and lipids, and secondary metabolites such as steroids, antibiotics, alkaloids, terpenoids etc can be isolated and analysed using TLC.

Plate 4 produces images of various leaf extracts using diethyl ether, ethyl acetate, methanol of M. jalapa. The results have predicted more number of compounds. Only few compounds are visualized as spots on TLC plates for ethyl acetate and methanol extracts. Hence there is more number of non polar compounds than polar compounds.

Based on present analysis, for diethyl ether plant extracts (Fig 1) we got 4 spots at 10% ethy acetate (RF: 0.22, 0.39. 0.65, 0.98), 4 spots at 30% ethyl acetate (RF: 0.04, 0.69, 0.89, 0.96), 1 spot at 50% ethyl acetate (RF: 0.11).

Based on present analysis, for ethyl acetate plant extracts (Fig 2), 2 spots at 10% ethy acetate(RF: 0.04, 0.08), 0 spots at 30% ethyl acetate, 1 spot at 50% ethyl acetate (RF: 0.06).

Based on present analysis, for methanol plant extracts (Fig 3), 0 spots at 10% ethy acetate, 1 spots at 30% ethyl acetate (RF:0.02), 0 spot at 50% ethyl acetate.

PLATE 4

Fig: 1. TLC plates with Diethyl ether extracts

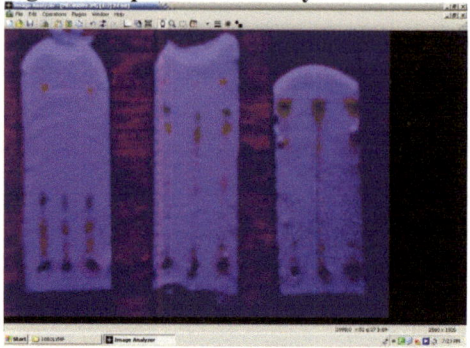

Fig: 2. TLC plates with Ethyl acetate extracts

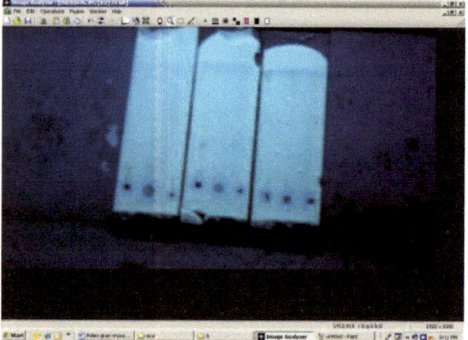

Fig: 3. TLC plates with Methanol extracts

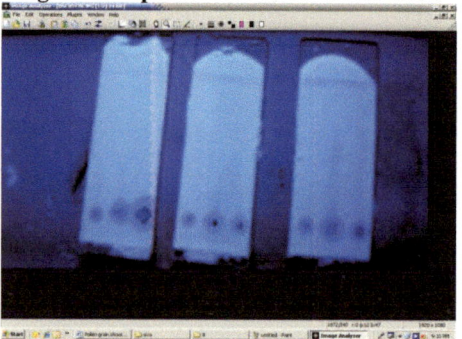

TABLE 1: Movement of Diethyl ether extract

Diethylether	ORANGE	PINK	YELLOW
10% ethyl acetate	Tot- 4.9 1.1, 1.9, 3.2, 4.8	Tot- 4.9 1.1, 1.9, 3.2, 4.8	Tot- 4.9 1.1, 1.9, 3.2, 4.8
30% ethyl acetate	Tot-5.1 0.2 3.5 4.1 4.9	Tot-5.1 0.2 3.5 4.1 4.9	Tot-5.1 0.2 3.5 4.1 4.9
50% ethyl acetate	Tot-4.6 0.5	Tot-4.6 0.5	Tot-4.6 0.5

Note: tot – movement of solvent, below movement of sample

TABLE 2: Movement of ethylacetate extract

Ethylacetate	ORANGE	PINK	YELLOW
10% ethyl acetate	Tot- 5.1 0.2, 0.4	Tot- 5.1 0.2, 0.4	Tot- 5.1 0.2, 0.4
30% ethyl acetate	Tot-5.0 -	Tot-5.0 -	Tot-5.0 -
50% ethyl acetate	Tot-4.9 0.3	Tot-4.9 0.3	Tot-4.9 -

TABLE 3: Movement of Methanol extract

Methanol	ORANGE	PINK	YELLOW
10% ethyl acetate	Tot- 5.0 -	Tot- 5.0 -	Tot- 5.0 -
30% ethyl acetate	Tot-4.9 0.1	Tot-4.9 0.1	Tot-4.9 0.1
50% ethyl acetate	Tot-4.8 -	Tot-4.8 -	Tot-4.8 -

ANTIMICROBIAL ACTIVITY

Plate 5 has shown the presence/absence of antimicrobial activity of *M.jalapa* leaf extracts against Bacteria and Fungi. Table 2 provided *in vitro* antibacterial activities of the extracts of *M.jalapa* and standard antibiotic (ampicillin). The solvent controls did not show any activity against the microorganisms used in this study. The methanol and ethyl acetate extracts did not showed activity against Gram-positive, Gram-negative bacteria or fungal organisms tested at 20μg/ml. The Diethylether extracts of all varieties showed the zone of inhibition against *E.coli* (9 mm), *S.aureus* (14 mm), *B. subtilis* (20 mm), *K. pneumoniae* (13mm), which is greater than the standard antibiotic used. The Diethylether extract did not showed the zone of inhibition against *A.niger*. Hence the results are provided that the Diethyl ether extracts of *M.jalapa* contain anti-pneumonic compounds. Ampicillin did not show any action action on Gram –ve bacteria and fungi. The Diethyl ether extract more zone of inhibition in Gram +ve compared with Gram –ve bacteria and fungi.

PLATE 5
Antimicrobial activity

Fig:1. *E. coli*

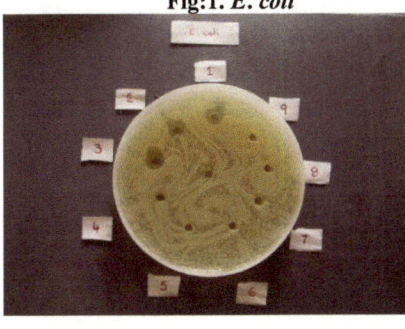

Fig: 2. *S.aureus*

Fig: 3. *B. subtilis*

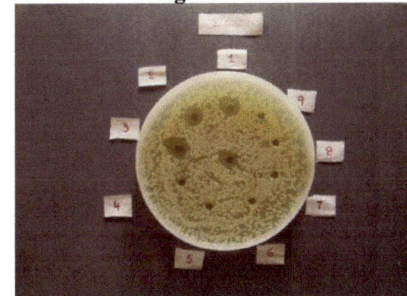

Fig: 4. *K.pneumonia*

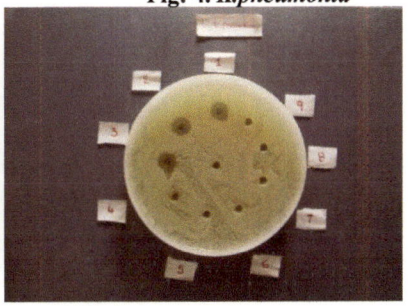

Fig: 5. *A. niger*

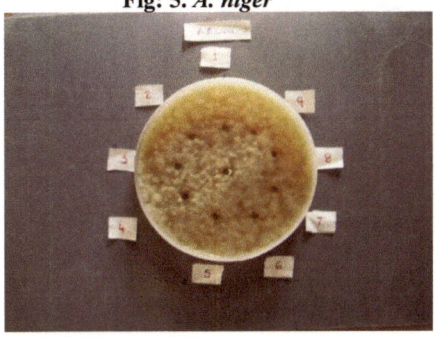

TABLE 4: Antimicrobial activity

Solvent	20µg/ml	E. coli (G-)	S. aureus (G+)	B. subtilis (G+)	K. pneumoniae (G-)	A. niger
Dietly ether extracts	Orange	9	14	20	13	---
	Pink	8	12	16	13	---
	Yellow	9	14	20	13	---
Ethyl acetate	Orange	---	---	---	---	---
	Pink	---	---	---	---	---
	Yellow	---	---	---	---	---
Methanol	Orange	---	---	---	---	---
	Pink	---	---	---	---	---
	Yellow	---	---	---	---	---
Ampicillin		---	13	13	---	---

PLATE 6
Bar graph for Antimicrobial activity

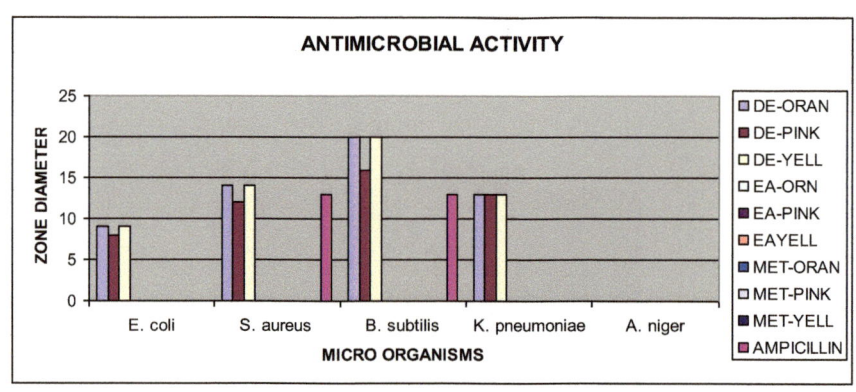

IMAGE ANALYSIS

Plate 7 has provided the outputs of the image analysis software Pixcavator which shows clarity of the image before and after adjusting the image. Figure shows the visualization of pollen grain and figure b provided the visualization of spots on TLC plate. Hence the software provided better visualizations after adjusting brightness, contrast, Hue and saturation.

Plate 8 provided the stained pollen grains and the photographed image at 100X of yellow, pink and orange colored varieties of *M.jalapa*.

Plate 9 to 12 shown image satisfying the current analysis settings. For each objects the following data is displayed:

- More number of dark than light objects,
- location (coordinates of the centroid = the center of mass of the object as a lamina of uniform density),
- size (the number of pixels = area),
- perimeter,
- roundness (above 80 for circles, less for everything else),
- intensity (max/min level of gray),- more number of toundness, ledder the intensity
- location of the center of mass (of the object as a lamina with density = gray level), -- not currently
- average contrast (saliency/size),

PLATE 7

Figure a

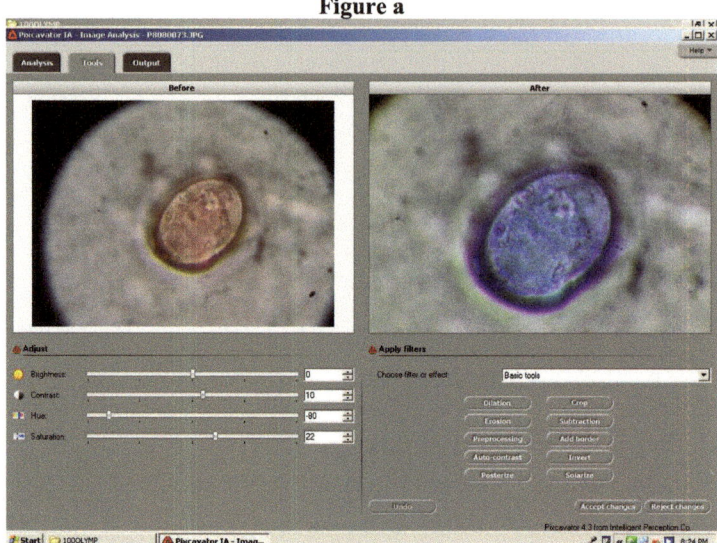

Figure b

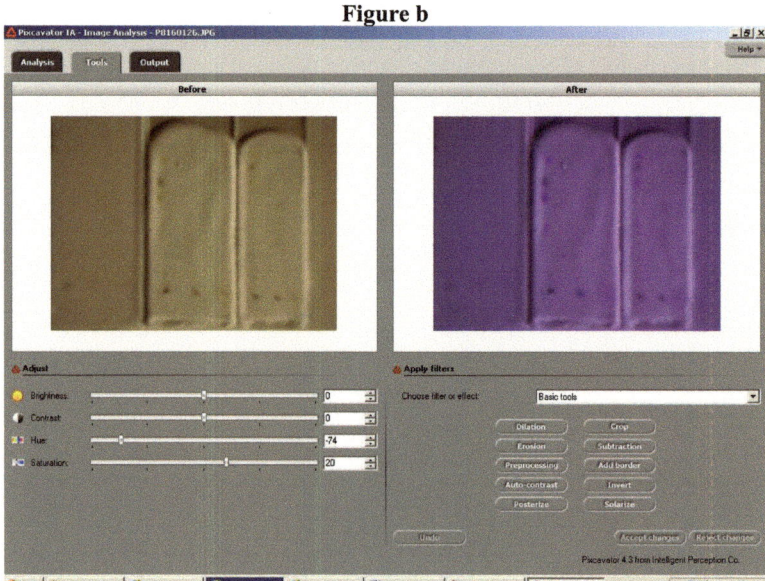

PLATE 8

Figure a : Slides of yellow, Pink and Orange stained plates

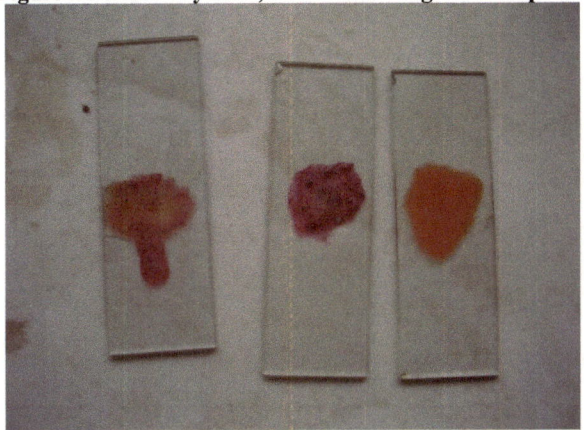

| Figure b: Yellow | Figure c: Pink | Figure d: orange |

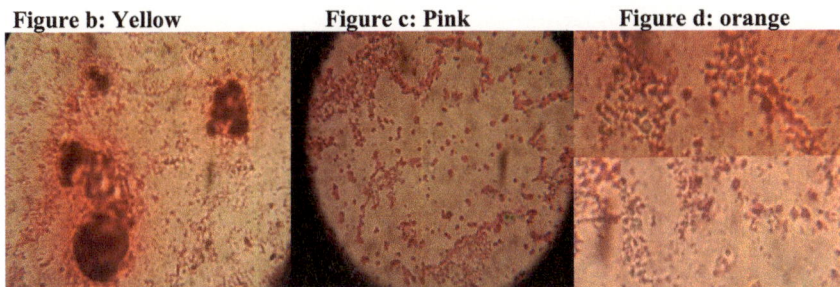

PLATE 9

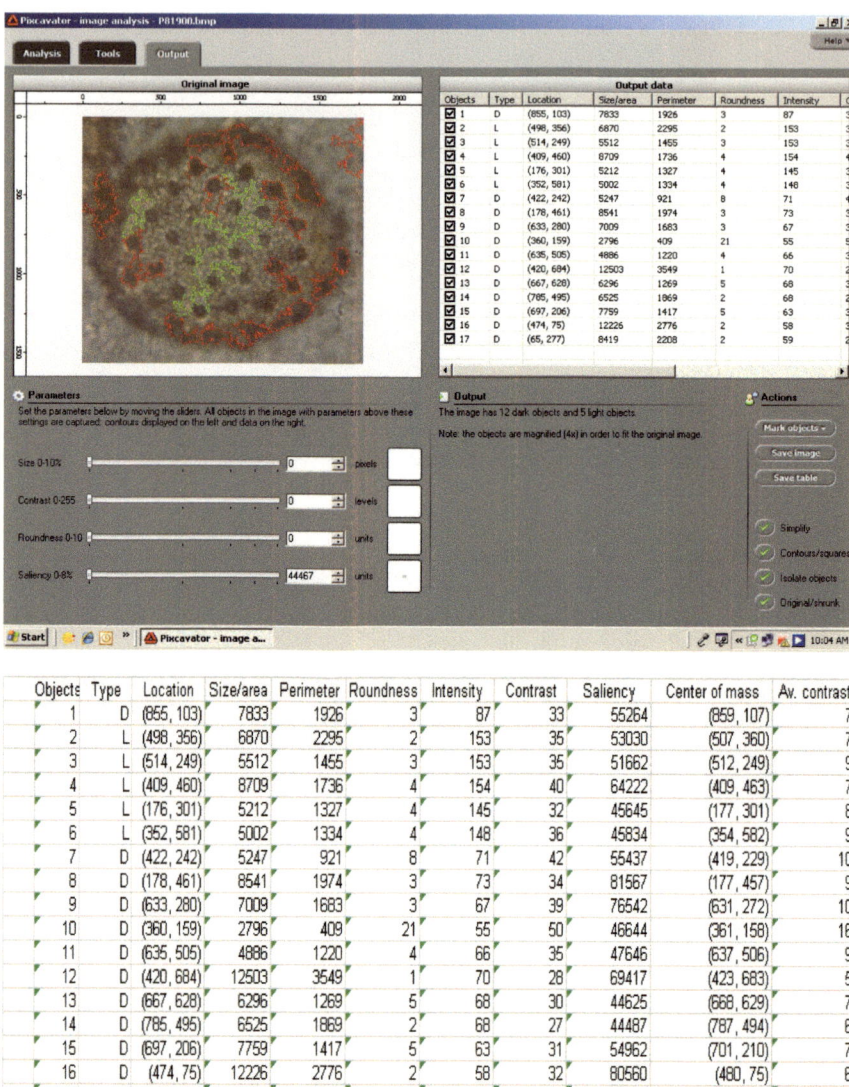

Objects	Type	Location	Size/area	Perimeter	Roundness	Intensity	Contrast	Saliency	Center of mass	Av. contrast
1	D	(855, 103)	7833	1926	3	87	33	55264	(859, 107)	7
2	L	(498, 356)	6870	2295	2	153	35	53030	(507, 360)	7
3	L	(514, 249)	5512	1455	3	153	35	51662	(512, 249)	9
4	L	(409, 460)	8709	1736	4	154	40	64222	(409, 463)	7
5	L	(176, 301)	5212	1327	4	145	32	45645	(177, 301)	8
6	L	(352, 581)	5002	1334	4	148	36	45834	(354, 582)	9
7	D	(422, 242)	5247	921	8	71	42	55437	(419, 229)	10
8	D	(178, 461)	8541	1974	3	73	34	81567	(177, 457)	9
9	D	(633, 280)	7009	1683	3	67	39	76542	(631, 272)	10
10	D	(360, 159)	2796	409	21	55	50	46644	(361, 158)	16
11	D	(635, 505)	4886	1220	4	66	35	47646	(637, 506)	9
12	D	(420, 684)	12503	3549	1	70	28	69417	(423, 683)	5
13	D	(667, 628)	6296	1269	5	68	30	44625	(668, 629)	7
14	D	(785, 495)	6525	1869	2	68	27	44487	(787, 494)	6
15	D	(697, 206)	7759	1417	5	63	31	54962	(701, 210)	7
16	D	(474, 75)	12226	2776	2	58	32	80560	(480, 75)	6
17	D	(65, 277)	8419	2208	2	59	28	52691	(64, 270)	6

PLATE 10

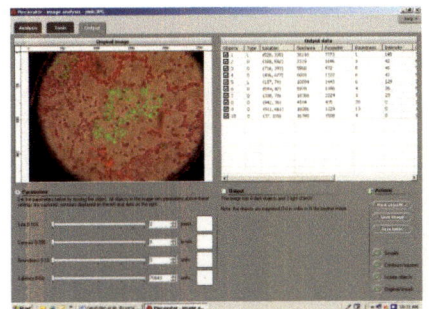 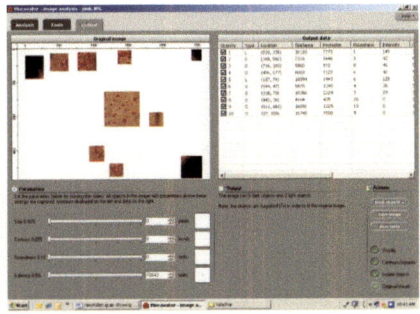

Objects	Type	Location	Size/area	Perimeter	Roundness	Intensity	Contrast	Saliency	Center of mass	Av. contrast
1	L	(528, 335)	38110	7773	1	145	31	182359	(524, 338)	4
2	D	(388, 582)	7319	1646	3	42	62	95819	(394, 579)	13
3	D	(716, 393)	5960	972	8	46	56	103160	(714, 390)	17
4	D	(496, 677)	6003	1122	6	42	56	79524	(489, 679)	13
5	L	(187, 74)	10094	1443	6	129	36	90519	(189, 77)	8
6	D	(544, 47)	5978	1398	4	36	52	82918	(547, 45)	13
7	D	(338, 70)	10366	2224	3	29	54	161630	(339, 70)	15
8	D	(942, 36)	4164	435	28	0	26	80064	(944, 32)	19
9	D	(911, 661)	16091	1229	13	0	11	94286	(918, 663)	5
10	D	(37, 100)	16740	1508	9	0	9	91959	(35, 83)	5

PLATE 11

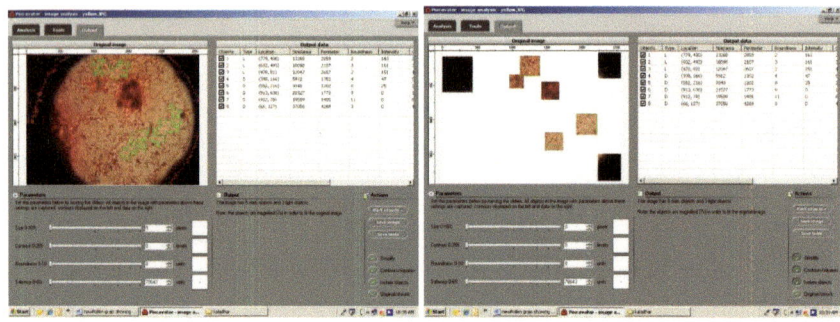

Objects	Type	Location	Size/area	Perimeter	Roundness	Intensity	Contrast	Saliency	Center of mass	Av. contrast
1	L	(779, 400)	13268	2859	2	163	29	85809	(774, 402)	6
2	L	(602, 493)	10098	2107	3	161	34	78672	(607, 486)	7
3	L	(470, 81)	12047	2607	2	151	40	99544	(472, 82)	8
4	D	(398, 166)	5912	1352	4	47	56	85946	(400, 165)	14
5	D	(582, 216)	9348	1202	8	25	27	79371	(582, 212)	8
6	D	(913, 636)	21527	1773	9	0	10	100699	(921, 640)	4
7	D	(912, 78)	19509	1485	11	0	6	87454	(916, 70)	4
8	D	(66, 127)	37056	4268	3	0	4	81371	(64, 109)	2

PLATE 12

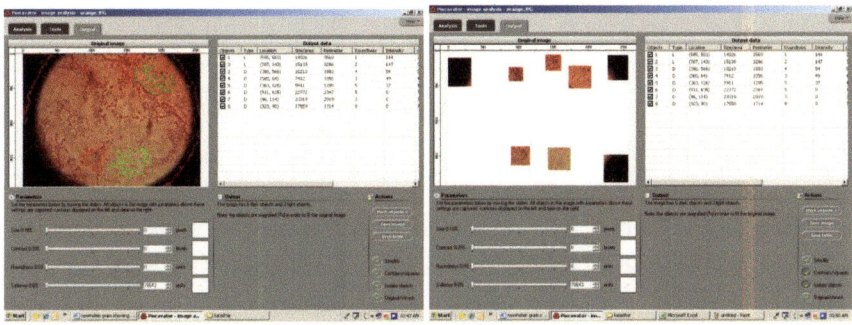

Objects	Type	Location	Size/area	Perimeter	Roundness	Intensity	Contrast	Saliency	Center of mass	Av. contrast
1	L	(595, 583)	14926	3569	1	144	22	80397	(602, 584)	5
2	L	(707, 143)	15139	3286	2	147	37	91522	(698, 145)	6
3	D	(386, 566)	10210	1803	4	54	55	91955	(392, 554)	9
4	D	(565, 64)	7412	1856	3	49	44	84009	(566, 63)	11
5	D	(363, 126)	5911	1205	5	37	49	86021	(363, 125)	14
6	D	(911, 635)	22772	2347	5	0	7	81356	(916, 638)	3
7	D	(46, 114)	23319	2919	3	0	7	90054	(44, 95)	3
8	D	(923, 90)	17559	1714	8	0	7	86389	(926, 75)	4

DISCUSSION AND SUMMARY

In recent years the preservation of local knowledge, the promotion of indigenous medical systems in primary health care, and the conservation of biodiversity have become even more of a concern to all scientists working at the interface of social and natural sciences but especially to ethno pharmacologists.

Further acquaintance with different ethnic groups has contributed to the development of research on natural products, to the increase in knowledge about the close relationship between the chemical structure of a certain compound and its biological properties, and to the understanding of the animal/insect-plant interrelation. For these reasons, medicinal plants are important substances for the study of their traditional uses through the verification of pharmacological effects and can be natural composite sources that act as new anti-infectious agents.[22] Thus, biologically active compounds present in plant products act as elicitors and induce resistance in host plants resulting in reduction of disease development. [23]

Silica plates were prepared by preparing slurry of silica powder in water in the ratio 1:2 (w/v). This slurry was poured onto the plates with the help of an applicator. The slurry was coated onto the glass plates. The plates were allowed to dry for 15-30 min. These plates were then heated in an oven at 100-120°C for 1-2 h to remove the moisture and to activate the adsorbent (silica gel) on the plate. Fractions were collected from the crude solvent extracts that were effective in reducing the virus concentration on the indicator plants of *N. glutinosa*. The solvent extracts were passed through a silica column (60-100 mesh) and 4 mL fractions were collected. The fractions were spotted onto a silica plate of 0.3 mm thickness. Presence of a single UV fluorescent band confirmed the purity of the fraction. Retardation factor (R_F) of each fraction was calculated using[24]:

$$R_{F=} \frac{\text{Distance travelled by the compound}}{\text{Distance travelled by the solvent front}}$$

Based on present analysis, for diethyl ether plant extracts we got 4 spots at 10% ethy acetate(RF: 0.22, 0.39. 0.65, 0.98), 4 spots at 30% ethyl acetate (RF: 0.04, 0.69, 0.89, 0.96), 1 spot at 50% ethyl acetate (RF: 0.11). Ethyl acetate plant extracts produced 2 spots at 10% ethy acetate(RF: 0.04, 0.08), 0 spots at 30% ethyl acetate, 1 spot at 50% ethyl acetate (RF:

0.06). Methanol plant extracts produced 0 spots at 10% ethy acetate, 1 spots at 30% ethyl acetate (RF:0.02), 0 spot at 50% ethyl acetate.

Mirabilis jalapa L., commonly known as 'four o'clock plant' produces a strong, sweet smelling fragrance after the flowers open at late afternoon. It is a well known ornamental plant as the flowers of different colours can be found simultaneously on the same plant or an individual flower can be splashed with different colours. The colour-changing phenomenon is one of the unique characteristics of *M. jalapa* as it can display flowers with different colour when it matures. Apart from its ornamental value, it has also earned its place in herbal medicine practices around the world. Its array of biological activities continues to support its use worldwide for control of viruses, fungi and yeast. [25]

According to R. NAIR et al 2005,[26] the antibacterial activities of *H. rosasinensis* and *M. jalapa* are shown at 40 mg/0.1 ml and neither aqueous nor methanolic extracts were able to inhibit any of the tested bacterial strains.

Most of the work is done on anti viral compounds. Hence scientists taught that there r no antibacterial components from *M.jalapa*. The present research provided that the activity is present if the compound solvent used is diethyl ether. The Diethylether extracts of all the three varieties showed the zone of inhibition against *E.coli* (9 mm), *S.aureus* (14 mm), *B. subtilis* (20 mm), *K. pneumoniae* (13mm), which is greater than the standard antibiotic used. The Diethylether extract did not showed the zone of inhibition against *A.niger*.

The results indicated that the antimicrobial activity against Gram-positive was more pronounced than against Gram-negative bacteria. The results obtained are in agreement with the work of Nair et al.,[27] Parkeh and Chanda,[28] and Encarnacion et al.[29] The differences may be attributed to the fact that the cell wall in Gram-positive bacteria consists of a single layer, whereas the Gram-negative bacteria it is a multilayer structure and is quite complex.

The ethanolic extract of the leaf of *Mirabilis jalapa* was tested by Oladunmoye et al., 2007 for antimicrobial activity against five pathogenic bacterial strains: *Escherichia coli, Staphylococcus aureus, Salmonella typhi, Bacillus cereus and Klebsiellapneumoniae.* Antagonistic activities of toxins from nine strains of bacteria and four fungi isolated from refuse dumps was also tested on the pathogens. The isolates producing the toxins were identified as *Pseudomonas* sp., *Acinetobacter* sp., *Cotynebacterium* sp., *Actinomyces* sp., *Clostridium* sp., *Bacillus* sp., *Shigella* sp., *Proteus* sp. *Enterobacter* sp., *Penicillium* sp., *Aspergillus flavus*, *Aspergillus niger* and *Aspergillus repens*. The agar ditch diffusion method was used was used for the in vitro antimicrobial bioassay. The leaf extract was found to have a higher antimicrobial efficacy than the toxins from the organisms as shown in the growth inhibition indices. The highest zone of inhibition of leaf extract was 13.0 mm and the least 4.0 mm. The highest zone of inhibition of bacteria isolates was 9.0 mm and the least 3.5 mm while the highest zone of inhibition of fungi isolates was 13.0 mm and the least 2.0 mm. Fungi isolates produce toxins with broad spectrum of activity. However the commercial antibiotics was found to be more potent than the toxins or the extract; But with narrower spectrum of activity. Phytochemical screening of the extract revealed the presence of tannins, saponins, alkaloids and cardiac glycosides. The toxins and the plant extract possess' antimicrobial activities comparable to conventional antibiotics; and can thus be a good source of agents for biocontrol and chemotherapy.[30]

The fate of plastid and mitochondrial nucleoids (pt and mt nucleoids) of*Triticum aestivum* was followed during the reproductive organ formation using fluorescence microscopy after staining with 4'6-diamidino-2-phenylindole (DAPI). This investigation showed a drastic morphological change of pt nucleoids during the differentiation of reproductive organs from the shoot apex. Dot-shaped pt nucleoids grew into ring-shaped ones, which divided into small pieces in the monocellular pollen grain, as observed in this plant's earlier stage of leaf development. During the development of mature pollen grain from monocellular pollen grain, pt and/or mt nucleoids disappeared through the division of the male generative cell of *T. aestivum*.[31]

The surface of pollen grain is having pores which are arranged in serial fashion at half posterior and axial another anterior. The posterior half has pores arranging from 1 to 5. The anterior portion is ready and ranges 1, 3, 5 on first half and another half. The anther is a 2 lobed structure attached to the stamen and is further subdividing into 7 sub lobes. The outer layer of the anther is thick, middle cellular and inner slimmer layers. The middle layer which is cellular may be represented as anther and the number is many

Image analyzer and pixcavator are the image analysis softwares provided good results about the picture clarity and various parameters such as objects, area, perimeter, roundness etc of the objects collected in microscopy taken by Olympus FE-115 MODEL 5 megepixel camara (2048X1536).

CONCLUSION

The present experimentation showed the good results in application of software's along with wet lab methodologies. Diethyl ether extract of M.jalapa can cure the diseases such as pneumonia, pus, wounds etc. The extracts contain large quantities of non polar compounds, which act against microbes. The architecture of anther is also complex with large number of sticky pores on surface containing 5 light and 12 dark types of images. Much research has to be conducted in future for adaptation of these plants in past and near future against bacteria, virus, fungi, birds, animals etc

BIBLIOGRAPHY

1. Farombi EO. African indigenous plants with chemotherapeutic potentials and biotechnological approach to the production of bioactive prophylactic agents. *African J Biotech* **2**: 662-671, 2003.

2. Stuffness M, Douros J. Current status of the NCI plant and animal product program. *J Nat Prod* **45**: 1-14, 1982.

3. Baker JT, Borris RP, Carte B, Cordell GA, Soejarto DD, Cragg GM, Gupta MP, Iwu MM, Madulid DR, Tyler VE. Natural product drug discovery and development: New perspective on international collaboration. *J. Nat. Prod.* 58: 1325-1357, 1995.

4. Reddy PS, Jamil K, Madhusudhan P. Antibacterial activity of isolates from Piper longum and Taxus baccata. *Pharmaceutical Biol* 39: 236-238, 2001.

5. Erdo¤rul OT. Antibacterial activities of some plant extracts used in folk medicine. *Pharmaceutical Biol* 40: 269-273, 2002.

6. Atefl DA, Erdo¤rul OT. Antimicrobial activities of various medicinal and commercial plant extracts. *Turk J Biol* 27: 157-162, 2003.

7. Maheshwari JK, Singh KK, Saha S. Ethnobotany of tribals of Mirzapur District, Uttar pradesh, Economic Botany Information Service, NBRI, Lucknow. 1986.

8. Rai MK. Ethnomedicinal studies of Chhindwara district (M.P.). I. Plants used in stomach disorders. *Indian Medicine.* 1: 1-5, 1989.

9. Negi KS, Tiwari JK, Gaur RD, Pant, K.C. Notes on ethnobotany of five districts of Garhwal Himalaya, Uttar pradesh, India. Ethnobotany 5: 73-81, 1993.

10. Taylor RSL, Manandhar NP, Hudson JB, Towers, G. H. N. Antiviral activities of Nepalese medicinal plants. **J Ethnopharmacol** 52:157-163, 1996.

11. Chung TH, Kim JC, Kim MK. Investigation of Korean plant extracts for potential phytotherapeutic agents against B-virus Hepatitis. *Phytotherapy Res* 9: 429-434, 1995.

12. Vlietinck AJ, Van Hoof L, Totté J. Screening of hundred Rwandese medicinal plants for antimicrobial and antiviral properties. *J. Ethnopaharmacol* 46: 31-47, 1995

13. Kusumoto IT, Nakabayashi T, Kida H. Screening of various plant extracts used in ayurvedic medicine for inhibitory effects on human immunodeficiency virus type 1 (HIV-1) protease. *Phytotherapy Res* 9: 180-184, 1995.

14. Dimayuga RE. Antimicrobial activity of medicinal plants from Baja California Sur/Mexico. *Pharmaceutical Biol* 36: 33-43, 1998.

15. Yang SW. Three new phenolic compounds from a manipulated plant cell culture, Mirabilis jalapa. *J Nat Prod* 64: 313-17, 2001.

16. Richard A. Niesenbaum, The effects of pollen load size and donor diversity on pollen performance, selective abortion, and progeny vigor in *Mirabilis jalapa* (Nyctaginaceae), *American Journal of Botany.* 86:261-268, 1999.

17. B P Cammue, M F De Bolle, F R Terras, P Proost, J Van Damme, S B Rees, J Vanderleyden and W F Broekaert. Isolation and characterization of a novel class of plant antimicrobial peptides form Mirabilis jalapa L. seeds. *The Journal of Biological Chemistry* 267: 2228-2233, 1992.

18. J Kataoka, N Habuka, M Furuno, M Miyano, Y Takanami, A Koiwai. DNA sequence of Mirabilis antiviral protein (MAP), a ribosome-inactivating protein with an antiviral property, from mirabilis jalapa L. and its expression in Escherichia coli. *The Journal of Biological Chemistry* 266: 8426-8430, 1991.

19. Kubo S., Ikeda T., Imaizumi S., Takanami Y., Mikami Y. A potent plant virus inhibitor found inMirabilis jalapa L. *Ann. Phytopathol. Soc. Jpn.* 56:481–487, 1990.

20. Ikeda T., Takanami Y., Imaizumi S., Matsumoto T.,   Mikami Y., Kubo S. Formation of anti-plant viral protein by Mirabilis jalapa L. cells in suspension culture. *Plant Cell Rep.* 6:216–218, 1987.

21. Jorge M. Vivanco, Maddalena Querci, Luis F. Salazar, Antiviral and Antiviroid Activity of MAP-Containing Extracts from *Mirabilis jalapa* Roots, *Plant Disease*, 83(12): Pages 1116-1121, 1999.

22. Viegas, C.; Bolzani, V.S. Os produtos naturais e a química medicinal moderna. *Quim. Nova*, 29: 326-337, 2006.

23. Verma, H.N., V.K. Baranwal, S. Srivastava, Antiviral Substances of Plant Origin. In: Plant Disease Control, Hadidi, A., R.K. Khetarpal and H. Koganezawa (Eds.), APS Press, St. Paul, Minnesota, 154-162, 1998.

24. N. Deepthi , K.N. Madhusudhan , A.C. Uday Shankar , H. Bhuvanendra Kumar, H.S. Prakash and H.S. Shetty , Effect of Plant Extracts and Acetone Precipitated Proteins from Six Medicinal Plants Against Tobamovirus Infection, *International journal of virology* 3(2): 80-87, 2007.

25. Anna Pick Kiong Ling, Kuok-Yid Tang, Jualang Azlan Gansau, Sobri Hussein , Induction and Maintenance of Callus from Leaf Explants of Mirabilis jalapa **L.**, *Medicinal and Aromatic Plant Science and Biotechnology* 3(1):42-47, 2009.

26. R. Nair, T. Kalariya, Sumitra Chanda, Antibacterial Activity of Some Selected Indian Medicinal Flora, *Turk J Biol,* 29: 41-47, 2005.

27. DSVGK Kaladhar, Siva kishore. Antimicrobial studies, Biochemical and image analysis in Mirabilis jalapa. *IJPT* 2(3): 683-693, 2010.

28. Parkeh J, Chanda S. Screening of aqueous and alcoholic extracts of some Indian medicinal plats for antibacterial activity. *Indian J Pharma Sci* 68:835-838, 2006

29. Encarnacion Dimayuga R, Virgen M, Ochoa N. Antimicrobial activity of medicinal plants from Baja California Sur (Mexico). *Pharm Biol* 36:33-34, 1998.

30. Oladunmoye, MK, Comparative Evaluation of Antimicrobial Activities of Leaf Extract of Mirabilis jalapa and Microbial Toxins on Some Pathogenic Bacteria. *Trends in Medical Research* 2(2): 108-112. 2007.

31. Shinichi Miyamura, Tsuneyoshi Kuroiwa, Toshiyuki Nagata. Disappearance of plastid and mitochondrial nucleoids during the formation of generative cells of higher plants revealed by fluorescence microscopy. *Protoplasma* 141(2-3): 149-159, 1987.